This report is part of a series of reports on technical rescue incidents across the United States. Technical rescue has become increasingly recognized as an important element in integrated emergency response. Technical rescue generally includes the following rescue disciplines: confined space rescue, rope rescue, trench/collapse rescue, ice/water rescue, and agricultural and industrial rescue. The intent of these reports is to share information about recent technical rescue incidents with rescuers across the country. The investigation reports, such as this one, provide detailed information about the magnitude and nature of the incidents themselves; how the response to the incidents was carried out and managed; the impact of these incidents on emergency responders and the emergency response systems in the community; and the lessons learned. The U.S. Fire Administration greatly appreciates the cooperation and information it has received from the fire service, county and state officials, and other emergency responders while preparing these reports.

This report was produced under contract EMW-94-C-4436. Any opinions, findings, conclusions, or recommendations expressed in this publication do not necessarily reflect the views of the U.S. Fire Administration or the Federal Emergency Management Agency.

Additional copies of this report can be ordered from the U.S. Fire Administration, 16825 South Seton Avenue, Emmitsburg, MD 21727.

The Derailment of the
Sunset Limited
Big Bayou Canot, Alabama
September 22, 1993

Local Contact: **Chief Edward A. Berger**
Fire Chief
City of Mobile Fire Department
701 St. Francis Street
Mobile, Alabama 36602

OVERVIEW

This report details the response of the Mobile, Alabama Fire Department (MFD) to the derailment of a passenger train in a remote section of the Big Bayou Canot, nine miles north of Mobile. Two-hundred and twenty people were on board the train when it derailed, caught fire, and, in the case of some cars, became submerged in the Bayou. One-hundred and seventy-three people survived the derailment of the *Sunset Limited.*

The majority of the "rescue" work was accomplished in under four hours. However, the incident was protracted because the MFD had to oversee recovery of the 47 deceased. This required sending divers into zero-visibility, subsurface, confined-space conditions. Furthermore, surface operations to support the body recovery required personnel to work in extreme heat and humidity for hours on end.

Use of the Incident Command System permitted the MFD to run a safe and efficient operation in the face of extremely adverse conditions. While the "technical" operations are not especially illuminating, this incident is extremely instructive in understanding the value of tight incident control that is afforded through the Incident Command System.

Key findings of this investigation are:

- The Incident Command System worked extremely well. It allowed control over resources that were spread over a wide geographic area.

- Mutual aid was critical to handle an incident of this magnitude. Having formal agreements in place before the need occurred was important.

- Preparation of documentation for the National Transportation Safety Board was very time-consuming. Good documentation, initiated early, is key.

- Communications were difficult. Cellular phones helped more than portable radios. Large quantities of back-up batteries were needed.

- Morgue duty was exceptionally emotionally draining. Personnel should be rotated through this assignment frequently.

- Rehabilitation of personnel was a top safety consideration. This was especially necessary because of the work environment and the high heat and humidity.

- Many of the extrication tools that are designed for automobiles are ineffective on trains. The AMTRAK *Emergency Evacuation Procedures* guide lists alternative means of emergency access.

Technical Rescue Incident Investigation Project Explained

The Technical Rescue Incident Investigation Project is an effort of the U.S. Fire Administration to document case studies of certain technical rescue incidents. The project seeks to produce documentation in a "lessons learned" format in order to provide local emergency responders, trainers, federal and state agencies, and other interested groups enhanced knowledge about technical rescue response and safety.

Applicability to *Sunset Limited* Derailment

The derailment, in the early hours of the morning, of a fully loaded passenger train into a remote waterway in an area without vehicular access might be thought to be the product of the imagination of a disaster exercise planner run wild. Unfortunately, this "worst-case scenario" did occur and tested the mettle of the MFD and emergency responders from several neighboring jurisdictions. While the actual technical operations were more directed towards recovery rather than rescue, the experiences of the incident managers and responders are instructive about the value of pre-incident planning, exercise, and the use of the Incident Command System.

Investigation Methodology

The research for this report was conducted through face-to-face interviews with many of the people who had incident command and operational responsibilities during the response to the derailment of the *Sunset Limited*. In addition to the interviews, documentation provided by the MFD and the National Transportation Safety Board (NTSB) was reviewed.[1] Finally, a site visit was made to view the scene of the derailment.

---

[1] The sections of this Technical Rescue Incident Investigation which deal with the sequence of events involved in the derailment of the *Sunset Limited* borrow heavily from the NTSB's report, NTSB/RAR-94101, "Derailment of AMTRAK Train No. 2 on the CSXT Big Bayou Carrot Bridge Near Mobile, Alabama, September 22, 1993." Sections of this report are taken verbatim from that report and are indicated with bold Helvetica print. All times from the NTSB report have been converted to military time.

Acknowledgements

This report could not have been written without the invaluable assistance of the Mobile Fire Department. Specific thanks are accorded to Chief Edward Berger for the hospitality he afforded and the resources he made available to complete the investigation. An equal measure of appreciation is due to Steve Huffman, the MFD Public Information Officer, who arranged and facilitated the numerous interviews necessary for the writing of this report. Finally, Boatswain's Mates Rooks and Tucker, of the U.S. Coast Guard Marine Safety Office/Mobile, provided marine transportation to and from the crash site.

I.   Introduction

Mobile (population 265,000) is a heavily industrial port on Mobile Bay, in southwest Alabama, 31 miles north of the Gulf of Mexico. The city has a central business district which lies adjacent to the port area. Low-, medium-, and upper-income residential areas surround the central business district, and many square miles of unusable swampland comprise an irregularly shaped "panhandle" which follows the Mobile River to the northeast. Because Mobile is a port and because paper production is a major industry, there is a large amount of hazardous material transported through and around the city.

MFD is the area's largest fire department. It serves an area of approximately 210 square miles. MFD has 432 personnel, 18 engine companies, five truck companies, seven ambulances, a fire boat, and several other specialized units. The department answers approximately 18,000 calls per year, of which about 12,000 are EMS-related.

On September 22, 1993 at 0253 hours, Amtrak Train No. 2, the *Sunset Limited*, with 220 people on board, derailed on a railroad bridge which crosses the 300-foot-wide Big Bayou Canot (pronounced "can-not"), in the Mobile Delta, approximately nine nautical miles from the mouth of the Mobile River. The ensuing crash caused the submersion of two passenger cars and fire to break out in a dorm-coach car, the baggage car, and a fuel cell. Forty-two passengers and five train crew members died, 111 passengers and crew members sustained injuries, and 62 people escaped without injury.

II.   Sequence of Events

On September 21, 1993, the *Sunset Limited*, proceeding from Los Angeles to Miami, was delayed 34 minutes for repairs at a normally scheduled stop in New Orleans. The train departed New Orleans at 2334 hours. It arrived in Mobile at 0230 hours, where it stopped for about three minutes to drop off and pick up passengers.

The following table describes the train as it left Mobile. Beginning with the lead locomotive, cars are listed in the position which they occupied in the consist.[2] AMTRAK floor plans are reproduced in Appendix A.

Train Car	Description
Locomotive 819	These Model F40PH diesel-electric locomotives deliver 3000 horsepower and operate on #2 diesel fuel (stored in a storage tank on the underside of the locomotive). They are equipped with a 480 Volt AC power generator in the rear of the engine room.
Locomotive 262	Same as Unit #819.
Locomotive 312	Same as Unit #819.
Baggage Car 1139	These cars transport passenger baggage. All baggage cars are equipped with a stretcher in addition to standard emergency equipment.
Crew Dorm 39908	These two-level cars have eight crew bunks and seating for 40 passengers on the upper level, and crew lounges and bathrooms on the lower level.
Superliner Coach 34083	These two-level cars seat 75 passengers.
Superliner Coach 34068	Same as Unit #34083.
Superliner Coach 34040	Same as Unit #34083.
Superliner Lounge 39973	These two-level cars provide casual seating, tables, and bar space for 68 passengers.

---

[2]The term "consist" refers to the string of cars which comprise the train.

Train Car	Description
Superliner Diner 38030	These two-level cars combine a dining area for 72 passengers on the upper level and a food preparation area on the lower level.
Superliner Sleeper 32067	These two-level cars have sleeping quarters and daytime private rooms. The upper level houses five duplex and ten economy roomettes; the lower level houses one large family bedroom, four economy roomettes, one large handicapped toilet, and five public toilets.

At 0055 hours, the towboat *Mauvilla* departed the National Marine Fleet, heading upriver from mile 5 on the Mobile River. A dense fog soon settled in, reducing visibility to almost nothing. Fog in the region is notorious for its thickness and the speed with which it descends on the river. The pilot of the *Mauvilla* stated that he could barely see the fore end of the load he was pushing. Although *the Mauvilla* was equipped with radar, the NTSB found that the pilot was not trained in its use.

In the heavy fog, and given the extremely winding nature of the Mobile River (see the map of the Mobile River Delta and the inset enlargement - Figures 1 and 2, below), it is understandable how the *Mauvilla* wound up in the Big Bayou Canot instead of the Mobile River when it attempted to find a place to tie off to shore and wait for the fog to lift.

At 0245 hours, in the Big Bayou Canot, the *Mauvilla* struck an object (later discovered to be the CSXT railroad bridge which crosses the Bayou at that point). The towboat pilot was unable to see the object which he hit, and he later stated that he thought he had run aground. Inspection of the bridge after the derailment showed the south end of the girder span had been displaced 38 inches to the west; this caused the east girder to protrude into the path of the *Sunset Limited.*

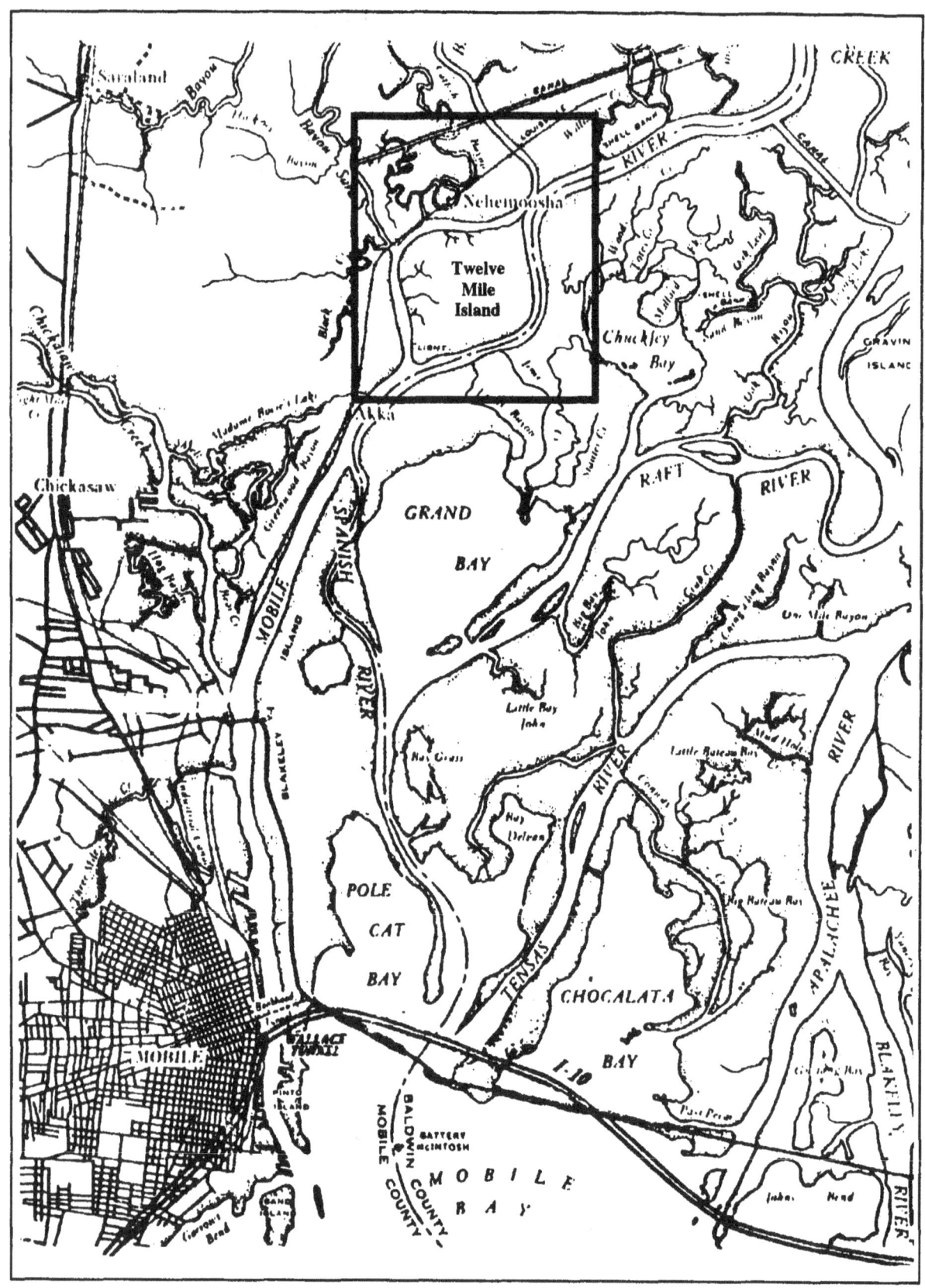

Figure 1 -- Map of the Mobile River Delta

4

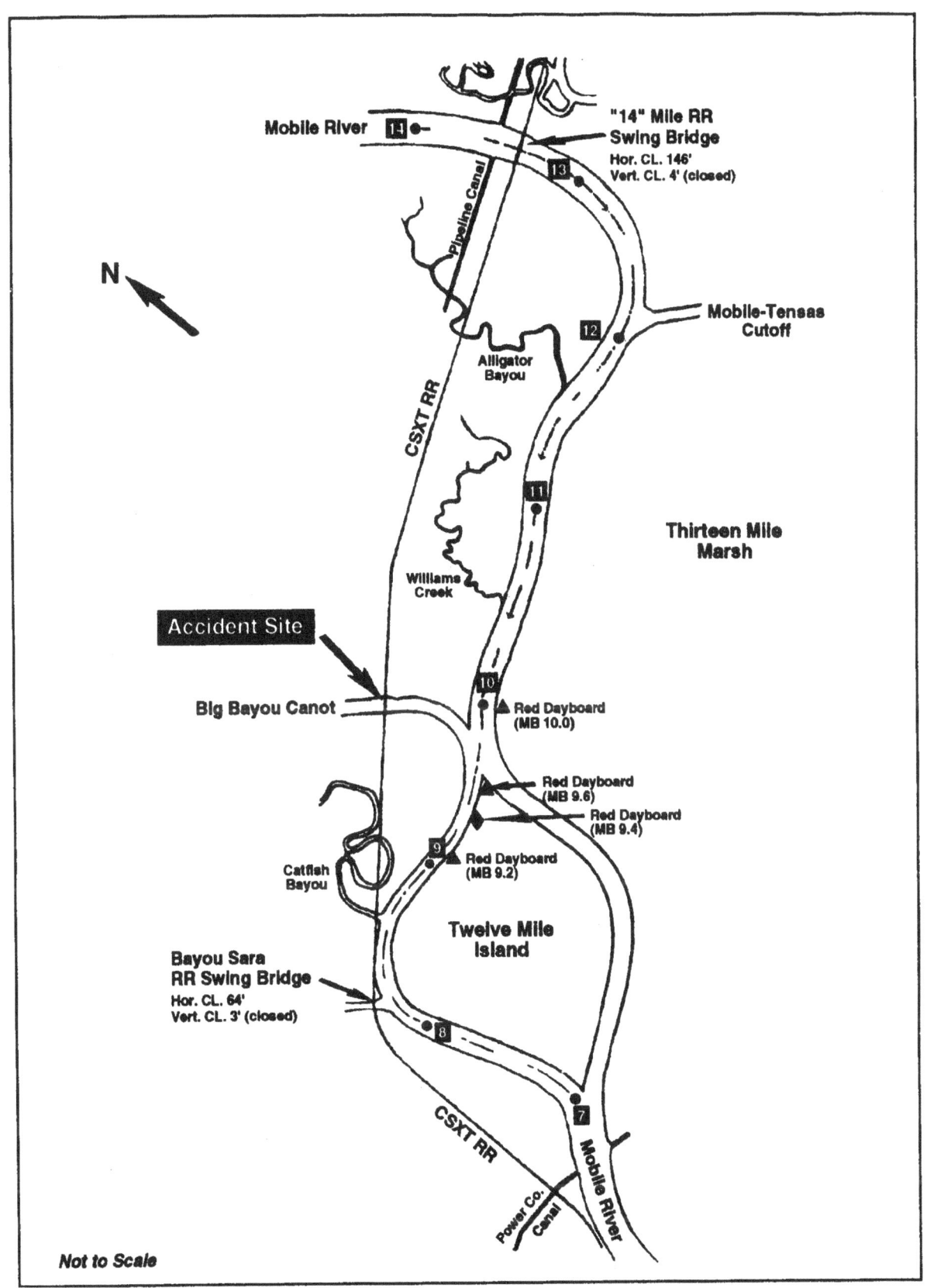

Figure 2 -- Inset enlargement showing the Big Bayou Canot

After departing Mobile, the *Sunset Limited* continued on its journey to Miami. There were 202 passengers and 18 crew on board; however, the exact passenger count would not be known until well after the derailment.[3] Traveling about 72 miles per hour, the *Sunset Limited* **struck the displaced bridge girder and derailed at milepost 656.7 about 0253 hours** (Figure 3 is an NTSB diagram of the bridge and the train.)

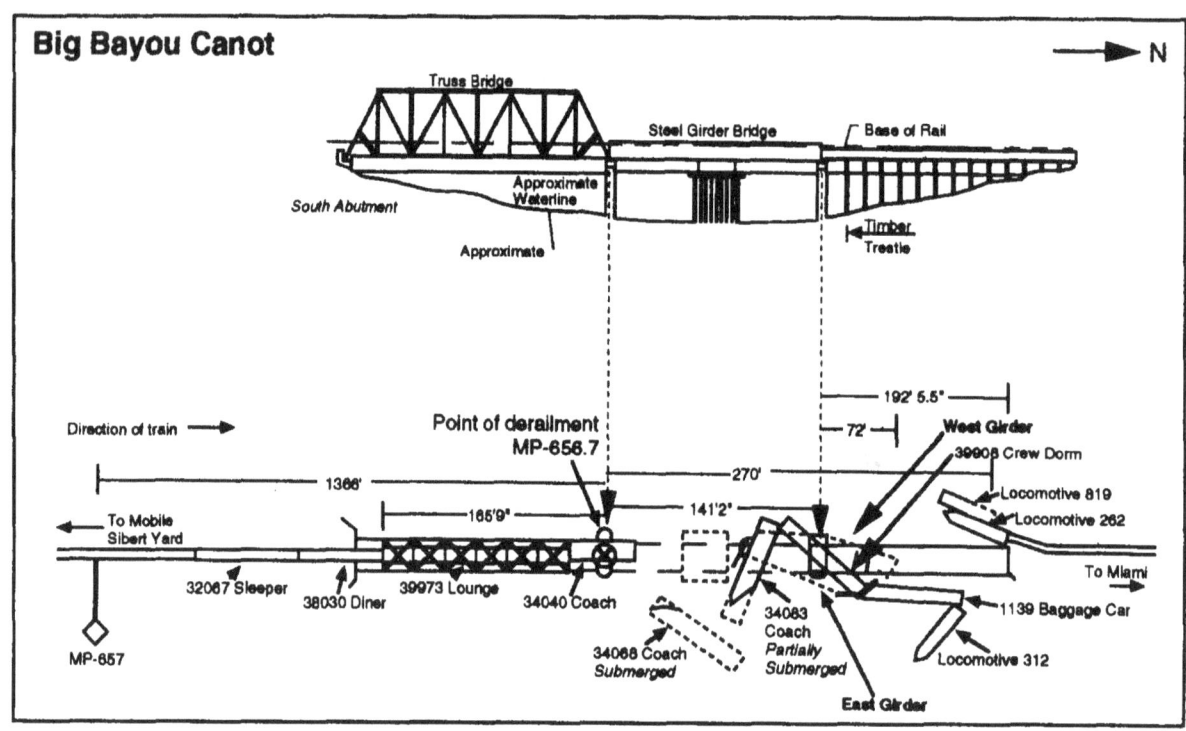

Figure 3 -- NSTB Diagram of Crash Site

Following the collision, the first seven cars of the train derailed into the waters of the Big Bayou Carrot. The lead locomotive, **819, was buried in about 46 feet of mud, and the part protruding above the embankment burned. The second [engine], 262, also burned. The fuel tank of the third [engine], 312, separated from it, and all equipment along the bottom of the unit below the frame was sheared off.**

---

[3]This was because AMTRAK policies did not require infants riding free of charge to be "ticketed." The problems this caused during the rescue/recovery operations will be discussed in a later section of this report.

Baggage car 1139 and dorm-coach 39908 . . . were gutted by fire, and parts of both cars sustained major structural damage. About half of coach 34083, which rested against the bridge after the accident, was submerged, and coach 34068 was almost totally submerged. The next four cars, coach 34040, lounge 39973, diner 38030, and sleeper 32067, remained on the bridge. All passenger cars were double-decker cars. (See Appendix A for applicable AMTRAK floor plans; see Figure 4 for a diagram of the wreckage).

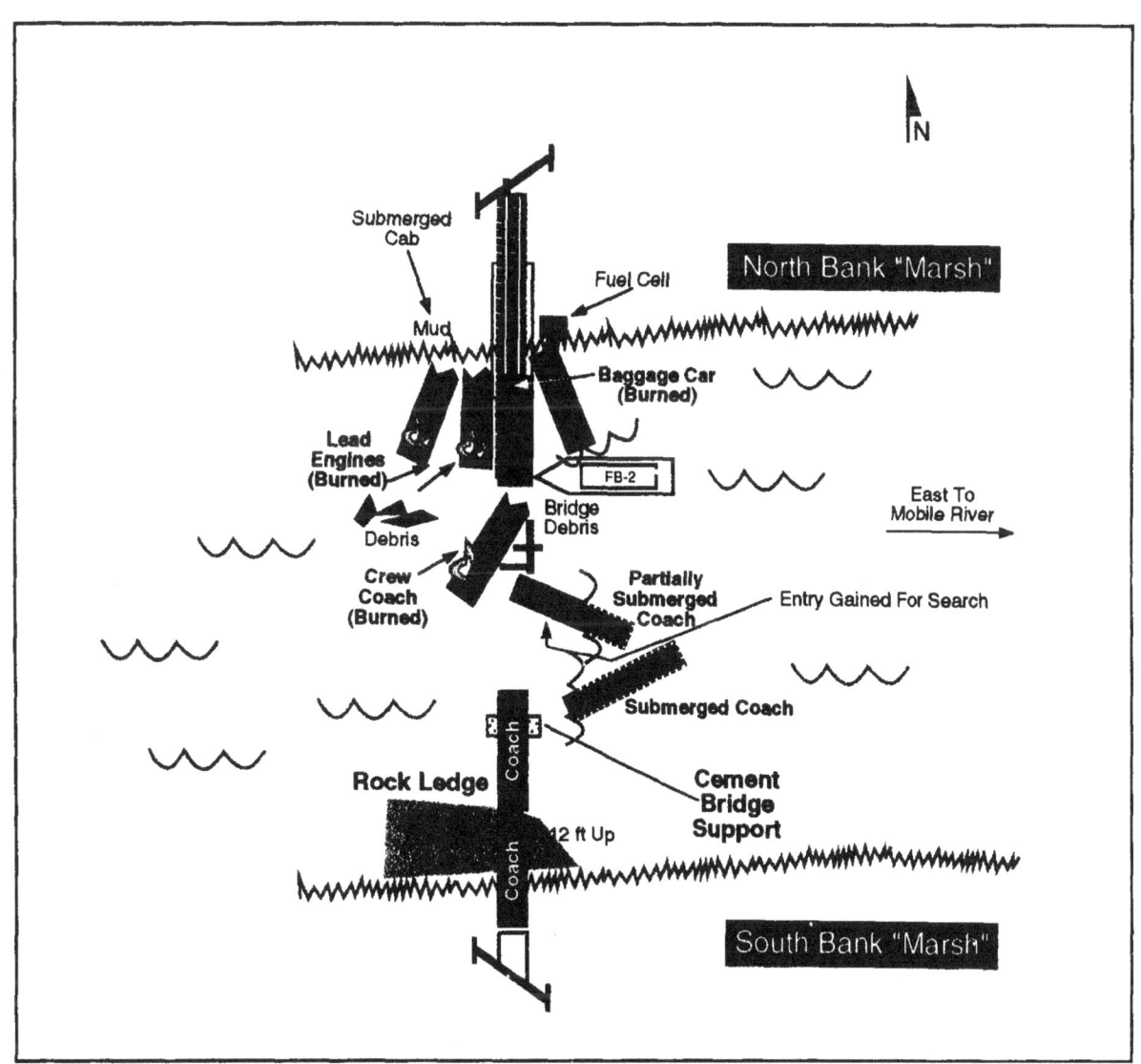

Figure 4 -- Diagram of Wreckage (based on a sketch by Capt. GA Foster)

Notification

[A]bout 0256 hours, [the] assistant conductor made a "Mayday, Mayday" transmission over the railroad-designated radio that was heard by CSXT train 579, waiting at [milepost] 660.4, whose crew repeated it to the yardmaster at the Sibert Yard, Mobile. Also about 0256 hours, the assistant terminal trainmaster at Sibert Yard heard [the *Sunset Limited*] transmitting Mayday over the radio. The yardmaster at Sibert Yard notified the train dispatcher in Jacksonville, Florida, at 0257 hours and the Mobile Police Department's 9-l-l operator about 0300 hours that [the train] had derailed. (See Appendix B for the transcript of the 9-1-1 notification.)

The bridge tender at the Mobile River Bridge and the engineer of train 579 also radioed the train dispatcher in Jacksonville about 0256 hours that [the *Sunset Limited*] was transmitting a Mayday call. Immediately thereafter, the train dispatcher tried to contact [the train] but was unsuccessful. Train 579's engineer advised the dispatcher that [the *Sunset Limited*] had derailed at the Mobile River Bridge, which is where the assistant conductor said he thought the train was when he made his Mayday call, and was on fire. The Mobile River Bridge is about 3.2 miles north of the actual accident site.

Between 0302 and 0305 hours, the Mobile Police Department's 9-l-l operator contacted the Mobile Fire Department and the Coast Guard. Police, fire, and Coast Guard personnel began notifying other emergency responders; more than 60 local departments eventually responded. [The *Sunset Limited's* On-Board Service (OBS)] supervisor, using a cellular telephone, called the Mobile 9-1-1 operator about 0305 hours and provided additional information about the accident location and what was taking place at the site. The OBS supervisor did not know the exact location of the derailment, however. For about 18 minutes - from 0302 to 0320 hours, confusion ensued as the Mobile, Saraland, and Chickasaw 9-l-l operators tried to locate the accident site. Exactly where [the train] had derailed was unclear, and no roads lead into the area, which is heavily wooded swampland.

First Response

At 0320 hours, 18 minutes after the initial call to 9-1-1, the MFD Fireboat, MFD Engines 1, 14, and 21, Truck 4, Rescue 14, the District 1 Chief, and the Medical Shift Commander were dispatched to the call. After the initial dispatch, the following agencies were notified: Mobile Police Department; Prichard Fire Department; Mobile County Sheriffs Department; the U.S. Coast Guard; the police and fire departments from Saraland, Chickasaw, Creola, and Mount Vernon; and the Mobile County Emergency Management Agency.

The initial dispatch location (Bayou Sara Drawbridge) was checked without finding anything. It was later determined that the location given the 9-1-1 operator was incorrect (see Appendix B for the 9-1-1 transcript).

The MFD Fireboat, the *Ramona Doyle,* departed its dock at 0320 hours. Navigation was clear until the fireboat reached the Cochran Bridge, at which point the visibility was reduced to zero by the same fog which had caused the pilot of the *Mauvilla* to become lost. From that point, the *Ramona Doyle* was forced to navigate by radar and slowed to about six knots.

While en route to the scene, Captain Randall Smith, the medical shift supervisor, called University of South Alabama Medical Center (the area's Level 1 trauma center). He requested that they put their helicopter, SouthFlite, on standby and that they notify all the other hospitals of a possible mass casualty incident. Captain Smith also notified Newman's Ambulance Service, the largest private ambulance service in the region at that time. He requested that they notify the other two private ambulance services and that they send units to stage at Saraland. Following the notification of the private ambulance services, Captain Smith contacted the communications center and ordered all off-duty paramedics called in. Finally, he gave orders for arriving units not to do CPR, not to perform endotracheal intubation, and not to give aggressive Advanced Cardiac Life Support, given the potential for large numbers of injured passengers.

Initial Actions -- Fireboat

The fireboat arrived on scene at about 0400 hours and confirmed the location of the crash as the Big Bayou Carrot. Heavy fire was showing in the baggage car, a crew dormitory car, two engines, and on the bridge itself. Twenty-eight survivors of the derailment had been plucked from the water by the crews of the Mauvilla and the Scott Pride.[4] After determining that no other people remained in the water, the Ramona Doyle began fighting the fires.

Scott Paper Command Established

Response personnel made several attempts to locate the crash site and a route for land-based vehicles to reach it. After determining that there was no road access, District One Chief V.E. Hall established command at the Scott Paper Company at 0420 hours. This site was chosen for a command post because it was the closest land location with both vehicular and rail access to the crash site.[5] For this reason, the Scott Paper Company also became the land-based staging area for the initial operations and the anticipated triage, treatment, and transportation site,

Accessing the Scene

Because it was an undetermined distance up the tracks from the staging area to the crash site (certainly further than walking distance), a "rescue train" was arranged to carry the initial wave of emergency responders to the scene. At 0455 hours, a three-car train carrying "Task

---

[4]The Scott Pride is a towboat which had been operating in the area and which responded to the U.S. Coast Guard's "Urgent Marine Information Broadcast" concerning the derailment (0324 hours).

[5]Beyond the paper mill, travel into the Mobile Delta could only be accomplished by boat or train.

Force One"[6] departed from the command post, with Captain Seth Peden from Engine 14 in charge.

Aboard the train, Captain Peden used the 30-minute travel time to organize and prepare the personnel with him. The first car (a dining car) was to be used to treat the injured (because there were tables upon which they could be placed). The middle car carried all the medical and rescue equipment. (All potentially useful medical and rescue equipment had been stripped from the vehicles and placed on the train.) The third car was to be used for the "walking wounded." Captain Peden stated that "it helped a lot to set up the operations and talk about the plan prior to our arrival on scene."

Initial Actions - Task Force One

Upon arrival, at 0525 hours, Task Force One commenced a primary search of the area and took a head count of the victims brought onto the train. The search was hampered because the only available light was from the fires burning on the bridge and in the train cars, and the fog which blanketed the area. Rescuers checked the cars on the bridge, the railbed, and the surrounding accessible areas. Once the primary search had been completed and all of the known patients had been taken on board, Captain Peden ordered the rescue train to depart. He sent two paramedics and two defibrillation-trained EMTs back on the train with 125 patients (mostly walking wounded). The total on-scene time for the first train was less then 20 minutes.

Once the train left, Task Force One personnel initiated a secondary search of the shore and the railbed areas. This search, however, failed to locate any more survivors.

---

[6]Task Force One comprised the crews from Engines 14, 1, and 21; Truck 4; and Rescues 14 and 3 (a "rescue" is an advanced life support ambulance), for a total of 17 responders.

Search Activities

The crew from Ladder 4 boarded a bass boat which had arrived at the scene of the wreck. They proceeded to the above-water portion of the partially submerged car, and searched without success for survivors. They then boarded the *Ramona Doyle* and assisted with fire suppression activities.

As soon as the fires had been extinguished and the train had cooled to a temperature which permitted entry, the crew dormitory was searched, and the badly burned bodies of two crew members were recovered.

At this point, it became clear that aside from those people who had already been transported from the scene, there would be no more survivors. Hence, a command decision was made to move from a "rescue mode" to a "recovery mode."

Triage, Treatment, and Transport of Patients

At approximately 0550 hours, the rescue train arrived at the Scott Paper triage area on the west side of the river. MFD EMS personnel handled the triage, treatment, and transport of the 125 patients from the rescue train. Some patients were taken by ambulance; the less serious were transported by city transit bus.

Word was received that 28 passengers who had been pulled from the Bayou were being transported downriver by a towboat to the Scott Paper Company docks which are on the east side of the river. The Director of Mobile County EMS System (MCEMSS), Dirk Young, was sent with eight MCEMSS ambulances to the Scott Paper docks to handle the patients on that side of the river. None of the 28 patients had sustained serious injuries, and they were all transported to the hospital.

The U.S. Coast Guard airlifted five passengers to Bates Field, and SouthFlite flew two patients directly from the crash site to the hospital.

The last patient was transported by 0800 hours from the Scott Paper treatment area on the west side of the river. MCEMSS personnel triaged, treated, and transported all 28 survivors on the east side by 0830 hours.

Body Recovery Efforts

About 0430 hours, SCUBA divers from the Mobile Police Department and the Mobile County Sheriffs Flotilla arrived on scene and began to prepare for recovery operations. A diving team from the U.S. Marine Corps also showed up - its personnel actually did the body removals. At 0545, divers began the process of removing the dead from the train cars. Later, divers from the state bridge inspections diving team arrived on scene. They had extensive experience diving in sub-optimal conditions, and the expertise to conduct the search for additional victims.

After fire suppression was completed, the *Ramona Doyle* became a floating forward command post and rehabilitation station. Due to an error at the dispatch center, MFD's Chief, Edward Berger, was not immediately notified of the derailment. By the time he had responded, the Scott Paper command post had already been established, and Chief Berger decided to proceed to the crash site. A boat from Daphne Volunteer Fire Department ferried him up to the scene. He arrived at the crash site at 0615 hours and took charge of all MFD activities at the scene.

A temporary morgue was established on a barge in the Bayou. The MFD Chaplain was called to the scene, and MFD Engine 14's personnel were placed in charge of coordinating the activities associated with preparing the bodies for transport back to a temporary morgue.

Train Removal/Recovery of Engineers

In the crash, the lead locomotive drove into the muddy banks of the north end of the Bayou Canot. The front 45 feet of it was buried in the mud at about a 30-degree angle. It was believed that three AMTRAK engineers were in the first locomotive.

The two submerged passenger cars were removed from the bayou by a barge crane during the afternoon hours of September 23. All victims had been removed from them by the divers. There was one major concern in removing the trains from the Bayou; a fiber optic cable which ran through the Bayou, alongside the CSX railroad tracks. This cable was a major communications link serving thousands of users, including computer networks and 9-1-1 systems, from Jacksonville, Florida to Los Angeles, California. Severing the fiber optic cable would mean a devastating loss of communications for the users of the cable. MFD personnel rigged a rope system in an effort to pull the cable out of the way. Despite a few tense moments, the MFD rigging system succeeded in saving the cable while the train cars were removed. The train cars were subsequently moved by barge downriver to the Alabama State Docks.

The next morning, after the other cars had been removed, a 700-ton barge crane pulled the 240-ton locomotive out of the mud. MFD personnel used a handline to wash the mud out of the engineer's compartment and removed the bodies of the three engineers. This brought the final death toll to 47.

By 2000 hours on Friday, September 24, the train cars and locomotives had been cleared and all the victims accounted for. The forward command post was shut down, and all personnel were released. The final MFD operations at the crash site were limited to retrieval of the train's "black box" and passenger effects by the crew of the *Ramona Doyle* the next day, Saturday, September 25.

III. Discussion

Several aspects of the rescue and recovery operations of the MFD at the Big Bayou Canot warrant further discussion at this point. In order to understand the critical role that the ICS played in this incident, it is important to get a sense of the obstacles that the MFD and the other responding agencies needed to overcome.

Response Difficulties

The most obvious problem was locating and gaining access to the crash site. Several factors combined to hamper emergency response to the derailment.

The crew members who would have known exactly where the train had derailed were in the lead locomotive and all three had all been killed on impact. The location reports for the crash site were inexact because there were no landmarks readily visible to the surviving train crew and the Mobile Delta swamp all looks the same (especially in the fog at night). Train personnel thought that they were farther away from Mobile than they actually were and initially reported that the train had derailed at the Mobile River bridge. Later reports described the crash site as Bayou Sara (about 1.5 miles closer to Mobile). For about 18 minutes, 9-1-1 center personnel tried to determine where the crash was and who to send based on the conflicting information they received.

Based on observations made during the clear day and on the descriptions of people interviewed who are familiar with the Mobile River and its tributaries, the area is confusing and inchoate - even in daylight. Given that it was a foggy night, responders would have been hard-pressed to arrive much faster even if they had had a verifiable location of the incident.

Although the MFD was dispatched in the initial group of emergency responders, the premise was that the MFD was responding on a mutual-aid basis to the Saraland Fire Department's response area. It was not until the following morning that it was determined that the crash site technically lay within the city limits of Mobile and the MFD had jurisdiction over the scene.

Finally, the few roads and paths into the Mobile Delta are spread out over a wide area, and are not well mapped or marked. After making a number of attempts to locate the crash site by land, the responding crews realized that road access was an uncertain means, and that the best chances for success would be via the railroad itself.

Scene Safety Considerations

The crash site itself was dangerous to rescuers and survivors alike.[7]

- The Bayou was 300 feet wide and about 20 feet deep where the submerged cars lay. There was an ever-present possibility of falling into the water.

- The railbed was elevated, and footing was unsure because the sides of the railbed were steep, gravel-covered embankments. In addition, railroad ties and other obstacles presented constant tripping hazards.

- The accessible portion of the bridge was obstructed by the train. The northern half of the bridge was mangled and on fire.

- The only light was from the fires which burned and the few spotlights on the *Ramona Doyle,* the *Mauvilla,* and the *Scott Pride.*

- The scene was 30 minutes by rail (and about the same by boat) from the nearest land access point, making emergency evacuation of any personnel who became

---

[7] Contrary to the rumors which circulated in the media, there was no danger present from snakes or alligators. Although both are known to inhabit the Mobile Delta, they are characteristically timid animals. The noise, fire, and commotion associated with the derailment and the response frightened away most the swamp's denizens.

injured more difficult than would otherwise be acceptable. This also posed problems for resupply operations.

- There was no landing zone big enough to accommodate a helicopter. Personnel and supplies that were airlifted to the scene had to be raised and lowered using a hoist.

- The wreckage of the *Sunset Limited* itself posed several hazards - glass, twisted metal, fire, and fuel were all present.

- During the day, temperatures reached the mid-nineties. This, combined with the high humidity and the extended exposure of rescuers, posed real problems for ensuring that personnel would not suffer heat prostration.

Dive Operations

The first-arriving rescuers made a limited search of the above-water portion of one of the cars in the Bayou. Beyond that attempt, the lack of lighting made further sub-surface attempts to search impossible. The "technical" aspects of the response were limited to providing underwater recovery operations inside the AMTRAK cars. This necessitated the use of highly trained divers to conduct a methodical search for and recovery of bodies. A commitment to safety dictated that precautions be taken to assure that no divers were injured performing body recoveries.

The water in the Bayou was a dark brown (one person interviewed described it as coffee-colored) from all the mud which had been stirred up. Visibility inside the train cars was less than one foot. This created an extremely hazardous environment in which to run recovery operations. Because of the cramped space and loose debris inside the cars, divers had to work alone, with a back-up rescue diver positioned at the entrance to each car in case the search diver got into trouble. The bodies that were located were brought out to a "shuttle" diver at the window of the car. The body would then be taken to the surface and placed on the morgue barge for identification.

Some of the divers used "Kirby Morgan" helmets. These specialized helmets have a tether which delivers surface-supplied breathing air, acts as a safety line, and facilitates communications with the surface and other divers. The use of these helmets permitted longer dive times and better communications. Some divers were also equipped with helmet-mounted cameras.

IV. Incident Command

Initial Establishment of Command

District One Chief Vernon Hall established command at the Scott Paper Company beneath the Cochran Bridge at 0403 hours. He immediately established Staging, Operations, and Medical Sectors.

Chief Hall directed District Chief Mike Byrd to assume responsibility for the Operations Sector. Chief Byrd was flown by helicopter to the *Ramona Doyle,* from where he directed the on-scene operations.

Scott Paper Command Post

At 0440 hours, Chief Hall transferred command to Deputy Chief Stephen Dean, who expanded the ICS organization to include Triage and Transportation Sectors. The Mobile County Emergency Management Agency's Communications Unit arrived on scene and was placed into service as the command post.

The Medical Sector Officer, Captain Randall Smith, prepared a patient triage and treatment area near the railroad tracks and the command post, and initiated staging of the many ambulances which would arrive at the staging area. Two helicopter landing zones were also established.

John Owen, Chief Hall's aide, was placed in charge of personnel accountability.

Decision to Move Command

Because of the necessity of moving operational and command personnel between the operations area and land, it became obvious that a command post/staging area with water access would be more advantageous. For this reason, at 0800 hours the command post was moved to the North Star Lumber Pier in the Port of Chickasaw. The move was completed and the new command post activated by 0845 hours.

Port of Chickasaw Command Post

The new command post offered several advantages. First, access to the site could be restricted. Second, it relieved the Scott Paper Company from having the congestion and logistical problems associated with having a large, 24-hour operation in front of its main gate. Third, there was water access for shuttling personnel to and from the incident site. Fourth, it was the receiving point for bodies and had ample room for use as an interim morgue.

The new command post also allowed the Public Information Officer to establish a press briefing area and to stage the press in between briefings.

Finally, the Port of Chickasaw had ample room for support services (such as food and lavatories) to be established.

Forward Command Post

Because the scene of the derailment was so far isolated from any accessible roads, a forward command post was established on the *Ramona Doyle*. This provided the Operations Sector an area from which to manage the scene while the Port of Chickasaw command post was used to coordinate the overall activities of the response.

The fireboat offered several inherent advantages for the operation. First, it offered shelter from the sun and had air conditioning and a refrigerator aboard (this was critical as the daytime temperatures hovered in the high-nineties). Second, it had radio communications which were not dependent on batteries (although the Big Bayou Carrot was in somewhat of a radio "dead zone"). Third, it offered a private place for command personnel to conference.

Difficulties of Command

The ICS had never been used on such a large scale, except in a drill. According to Deputy Chief Dean, the MFD uses the Incident Command System on a daily basis, but this was by far the largest incident the MFD had ever run. Chief Dean attributes the ability to use the ICS in such extreme circumstances to the department's constant exposure to ICS in the course of routine operations. Most others agreed that familiarity with the ICS was the main reason that incident management was not overly problematic.

An additional difficulty was that because of the location of the crash and the darkness, it was impossible to be sure exactly who had jurisdiction over the incident. As stated earlier, the MFD had been dispatched on the assumption that it was providing assistance to Saraland Fire Department or the County Sheriff. Institution of the ICS by the MFD was not done in an effort to gain command of the incident, but rather as a standard operating practice use by the MFD to manage incidents of all types.

Chiefs Hall and Berger were proceeding on the rationale that the incident was in Mobile County and that the Sheriff had jurisdiction. It was not until midday that it was actually determined that the incident was in the "panhandle" portion of the City of Mobile which extends along the railroad tracks well into the Mobile Delta. According to Chief Berger, he "eased into command" about noon the next day.

The MFD was the only agency which had been trained in and used ICS. This meant other agencies which responded to the derailment needed to be brought into the ICS organization without their having a proper understanding of how the system worked. Accordingly, the ICS organization remained staffed by MFD personnel but utilized the personnel and resources of the other agencies to complete incident objectives. While this approach worked reasonably well, most agreed that incident command would have been easier had officers from other agencies been able to dovetail into the ICS structure.

Number of Victims

"Not knowing" the exact number of victims was a source of considerable stress for the incident command staff. A number of rescue personnel cited the inability to determine consistent and reliable numbers of people on board as a major problem.

This problem arose because AMTRAK ticketing policies in place at the time of the derailment did not require that certain children traveling with revenue passengers or employees be ticketed.[8] This meant that they did not appear on the passenger manifest. Accordingly, there was no way to know when all the people had been accounted for. One responder was assigned the full-time task of coordinating the passenger count with the AMTRAK liaison. It was not until well into the recovery effort that AMTRAK was able to provide a comprehensive passenger list.

---

[8] AMTRAK is presently working to implement an on-board, computerized system which will allow complete and accurate passenger lists to be obtained in the event of a similar emergency. This system is projected to be available by sometime in 1996.

ICS Organizational Chart

Figure 5 shows the fully expanded ICS structure used for the incident.

Interagency Liaison/Unified Command Structure

Deputy Chief Dean was the Incident Commander. There was a high level of interagency cooperation with representatives of other agencies assigned to liaison positions at the command post.

Officials from each of the responding agencies as well as representatives from AMTRAK and CSX were on site at the Port of Chickasaw command post. The Operations Sector made frequent use of the technical expertise of AMTRAK's Chief Mechanic, especially during the task of lifting the cars and locomotives.

Media Liaison

Control of the media was a major task. By noon Thursday, there were over 300 news reporters and 75 tractor-trailer TV news trucks which created a line of parked vehicles over 1/4-mile long on the access road to the Port of Chickasaw command post.

Working together, the public information officers from the Mobile Fire and Police Departments, Steve Huffman and Tom Jennings, ensured that the media was kept from endangering themselves or interfering with the operations. The two held hourly briefings near the command post to keep the media fully informed of developments and supplied with accurate information that had been approved for release.

Sunset Limited Derailment
ICS Organization

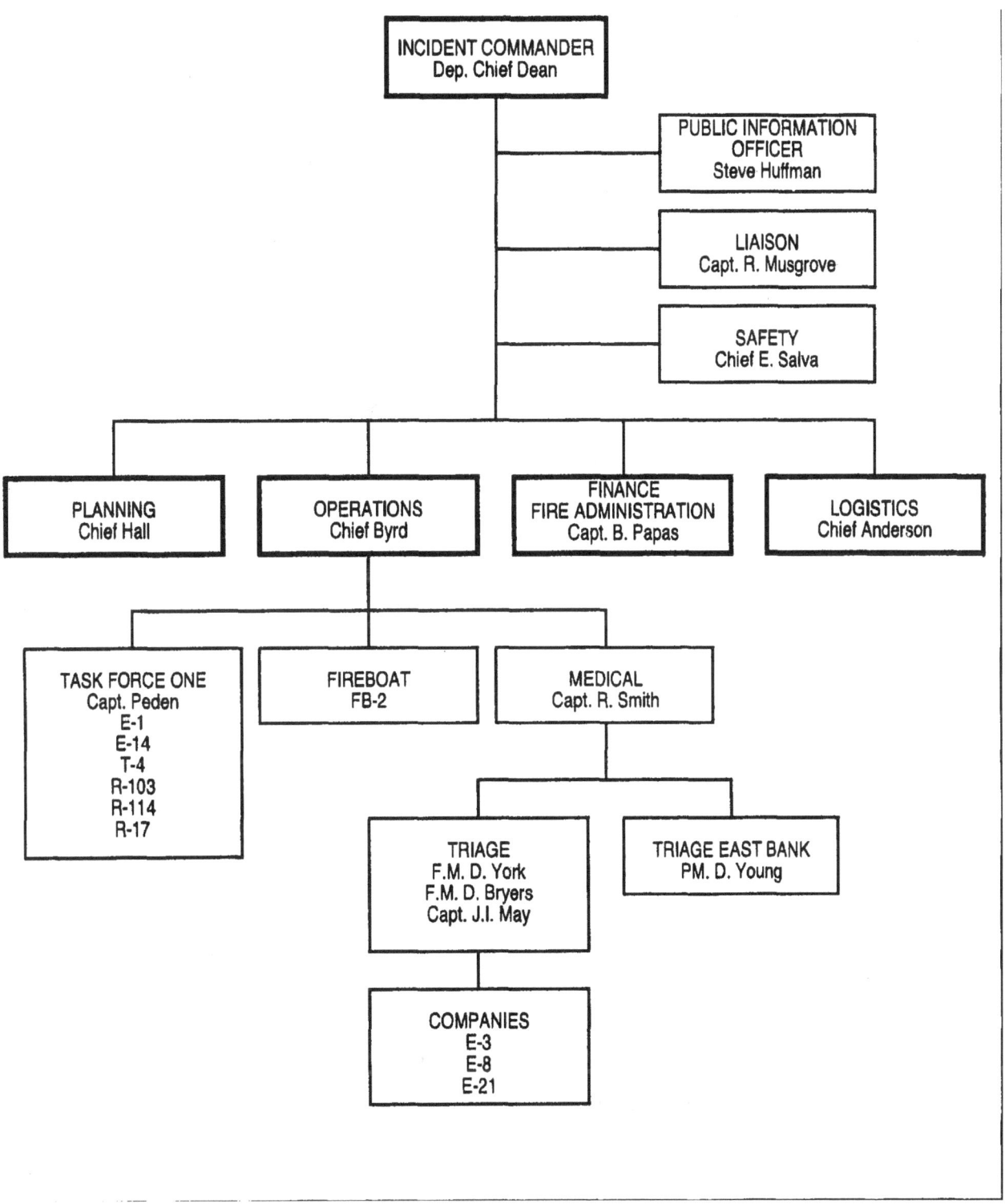

Figure 5 -- ICS Organization at peak expansion

The media was kept in a media staging area at the Port of Chickasaw command post. Media personnel were taken by boat in groups of ten to the crash site. To ensure fairness, the groups were chosen by lottery, with an even representation from print, TV, and radio outlets. All the media eventually did get to visit the scene.

Media representatives were afforded access to the portable latrines set up at the Port of Chickasaw and the food and drinks provided by the Salvation Army and the Red Cross. The media was asked to make contributions to a food fund. This was used to help the Salvation Army and Red Cross offset their expenses. This was not viewed as unreasonable by the media (who had large budgets to be there anyway). In retrospect, some responders thought that separating the media food and latrine facilities from those for emergency personnel might have been beneficial.

An additional consideration was the number of media helicopters which were circling the area. In order to avoid potential mid-air collisions and reduce the noise which was associated with the helicopters, Chief Dean had the air space in the area declared off-limits to non-essential aircraft by the Federal Aviation Administration.

Morgue Operations

Two morgues were established - one on a barge in the Bayou and one at the Port of Chickasaw command post. After initial recovery, the bodies would be lifted onto the barge, where they were placed in body bags. When a sufficient number of bodies were ready for transport, they were taken by boat to the Port of Chickasaw dock.

At the second morgue, the Identification Section of the Mobile Police Department photographed and made tentative identification of the bodies. Personal effects were secured by the police department, and the bodies were handed over to the Medical Examiner's Office for transportation to Mobile.

The crew from Engine 14, which had been on the initial rescue train with Task Force One, was initially assigned to handle the on-scene morgue. By all accounts, this crew worked the longest at the morgue. In retrospect, it was felt that a more frequent rotation of personnel through morgue duty would have relieved some of the emotional stress which morgue duty placed on those who performed it.

V. Major Findings

The following are some recurrent themes that emerged in the interviews:

ICS Saved the Day

By all accounts, ICS is credited as providing an organizational framework which created a systematic approach to handling this very chaotic scene. All MFD personnel interviewed indicated that the ICS worked the way it was intended to work.

The Mobile Police Department later commented on the functionality of the ICS and how well organized the MFD was. Since the incident, all public safety agencies in the area have adopted ICS and sought ICS training. The MFD has provided ICS training to a number of the area's public safety agencies.

Earlier Disaster Drill Was Helpful

On June 17, 1993, MFD personnel and equipment engaged in a disaster drill in conjunction with Mobile County Emergency Management Agency and Mobile County EMS System. The exercise tested the response to an airplane crash into Mobile Bay and simulated 90 survivors and 57 victims.

Many rescuers commented that the drill had been beneficial in refreshing their familiarity with disaster operations and that they felt that their ability to handle the train derailment had been enhanced by participation in the drill.

Rescue Operations Were Limited in Scope

Because the derailment occurred when most people were sleeping, there was no way for most of the passengers who died to escape their cabins once the cars became submerged in the murky water of the Bayou. Accordingly, by the time that the first rescuers arrived on scene, those who had been able to escape had been moved to safety. The "walking wounded" were the only survivors.

With the exception of the entry into the above-water portion of the partially submerged car, a fuller "search and rescue" effort was not feasible. For many rescuers, the frustration of "not being able to do anything" was the most traumatic aspect of the two-day operation.

The Limited Numbers of Seriously Injured Helped

Most rescuers agreed that the relative speed with which people were cleared from the patient receiving areas was due, in part, to the fact that so few people had serious injuries. Had there been greater numbers of injured, the EMS function would have been under much more pressure. Nonetheless, everyone interviewed felt that the Medical Sector had been properly established and was prepared to handle large numbers of seriously injured patients, had it been necessary.

Critical Incident Stress Management Played an Important Role

According to Chief Edward Berger, "[The wreck of the *Sunset Limited*] was the most traumatic incident any of us had ever been on." In recognizing and responding to that concern, Critical Incident Stress Management (CISM) was instituted to help responders deal with the emotional impact of the tragedy. Assistant Chief Pitt was assigned to oversee the provision of CISM.

At the time of the incident, only two of MFD's personnel had been formally trained in CISM.

Emergency responders were monitored on-scene for signs of traumatic stress; however, only one responder needed to be removed from the scene. All responders who continued to work on the incident underwent CISM.

Two phases were used. First, all responders were debriefed as they left the scene. Next, a formal debriefing was done on three dates in the week following the crash. All MFD personnel were required to attend one of the sessions. All personnel were paid for their time in debriefing sessions, thus facilitating acceptance and participation. In addition, the sessions were opened to personnel from any of the agencies involved.

The CISM activities were viewed as extremely successful. No MFD employee has been referred by the department to the Employee Assistance Program (although some may have sought help of their own volition). Only one MFD employee has subsequently left the department for reasons connected with the incident. No other individuals have reported major problems.

Preparedness for Future Responses

As mentioned earlier, agencies which had been involved in the response to the *Sunset Limited* derailment have subsequently decided to adopt the use of the Incident Command System. There is widespread belief that this incident demonstrated both the need for, and efficacy of, the ICS.

Subsequent to the incident, Mobile County EMS System recognized the need to have a large quantity of extra EMS equipment available for rapid deployment in future major emergencies. Through the efforts of its Director, Dirk Young, MCEMSS has purchased a trailer which is equipped with backboards and disaster supplies. Should large amounts of medical equipment ever be needed again, the trailer can be attached to the supervisor's vehicle or an ambulance and brought to the scene. This will help avoid the necessity of stripping ambulances for their equipment cache.

The MFD's communications system has been upgraded to an 800 mega-Hertz trunking system, and cellular phones have been purchased for use by ambulances, the hazardous materials team, select command staff, and the fireboat. By all accounts, the radio "dead space" problem appears solved. In addition, spare batteries for all communications equipment have been acquired.

VI. Lessons Learned

The following are lessons learned from this incident, as suggested by the personnel who were involved in its mitigation.

<u>Incident Management</u>

- It is essential to get an accurate count of the passengers. Without knowing how many people were on board, it was impossible to make a decision about when to discontinue searching for survivors or recovering victims.

- The use of the Incident Command System was what differentiated the fire department from all the other agencies which responded to the derailment. All agencies with emergency response duties should receive formal training in the use of the ICS, should use it on a daily basis on all incidents regardless of size, and should drill with other local agencies on its use. This will ensure that all personnel are comfortable and familiar with operating under the ICS when a major emergency occurs.

- The MFD needed to assure response capability to the rest of the city while it was handling the derailment. Incidents of this size quickly outpace the capabilities of even the largest departments. Without a formal mutual aid system in place, there would have been no way to get adequate numbers of personnel to the scene and ensure that additional calls were answered. The need for formal mutual aid agreements is especially applicable for small departments. This is demonstrated by the fact that the MFD, which has many more resources than other departments in the area, used mutual aid to effect the initial response.

- Work out bugs in the response system by drilling frequently with all other agencies which would be involved. Don't wait until the real thing to find out what works and what doesn't work.

- Initiate complete and accurate documentation early in the incident. On big incidents such as this, post-event investigations will be time-consuming and expensive. Proper documentation of the actions of the departments involved as well as the scene conditions will shorten the amount of time that needs to be spent reconstructing the paperwork afterwards.

- Coordination with other agencies is both helpful and problematic. Use the resources of other agencies to your advantage, but do not allow interpersonal issues to interfere with the mitigation of the incident. Work around problems and ensure that they receive adequate attention during the post-incident analysis.

Communications

- Good communications equipment is essential. After-action reports indicate that inter-agency frequency coordination was problematic. Personnel require a common channel for inter-jurisdictional or inter-agency communications. Radio dead zones need to be identified and written into pre-incident plans. Cellular telephones can provide a viable alternative means of communications.

- Have back-up batteries fully charged and ready to go at all times. Nickel-cadmium batteries should be "exercised" to prevent them developing a "memory." It is a good idea to have a bank of portable rapid-charge battery chargers and spare batteries available.

- This type of incident causes a heavy reliance on quality communications center personnel. Departments should ensure that they have adequate supervision in dispatch centers, that the personnel assigned there are well trained, and that they are included in all disaster exercises. Dispatch should be viewed as an integral part of the response team.

- While the initial responders assumed that it was too early in the morning for the incident to have been a drill, some expressed concern that some might have thought that they were being sent on a drill, had the incident occurred during the day. Communications policies should be to have all drills "dispatched" and run on a channel other than the main communications channel. In addition, an identifier should be used during the initial dispatch to ensure that responders understand that the incident is an exercise. This could be as simple as saying, *"ATTENTION, THE FOLLOWING IS A DRILL"* prior to, and after, the dispatch. This holds especially true for those agencies lacking alternate communications channels.

- Ground communications were hampered by the helicopters operating in the area. Minimize the number of unnecessary aircraft by implementing an airspace restriction around the incident. This can be done by contacting the closest air traffic control center (the phone number should be included in your agency's emergency operations plan).

Personnel Safety

- Personnel working on or around water must wear appropriate personal flotation devices (PFDs).

- Rotate crews and pace activities to avoid premature exhaustion. Don't let them get burned out in the initial stages of an incident - especially if it is likely to be an extended incident.

- Relieve the initial response personnel and remove them from the immediate operating area as soon as feasible. Critical Incident Stress Management (CISM) is an absolute must! The MFD personnel who were involved in the incident report that their participation in CISM was of significant benefit to them.

- Rehabilitation, although difficult to establish in the tight operational environment of the scene, was an absolute necessity. Given the high temperatures, constant exposure to the sun, and dangerous working conditions, it was critical to ensure that personnel were rehabilitated on a regular, formal basis. Use a large tarp or parachute cloth to set up a shaded area for rehabilitation.

- Personnel should be rotated through various jobs on an extended incident to prevent them from being unduly subjected to too much of any particular sight, sound, smell, etc. Ensure that this rotation occurs even if they are being adequately rehabilitated with rest periods, fluids, and nourishment.

- Personnel must receive ample food and liquids. This will prevent dehydration (which is the most likely source of fatigue) and loss of energy. Care must be taken to ensure that responders do eat and drink as they may have a tendency to ignore sustenance in favor of continuing to work. Personnel must not eat on scene or prior to thoroughly washing their hands.

In swampy areas, have ample supplies of mosquito repellent on hand. When working in the sun, make sure that sunblock is available to all personnel.

Logistics

- Resupply of food, water, and other supplies to the scene was difficult. Several hours often elapsed between the time that a need for supplies was recognized and the time those supplies could be delivered to the scene. This highlights the need for a planned logistics component within the ICS organization and for the early recognition of such needs.

- Don't underestimate the difficulty or importance of logistics. The MFD considered this an important enough function to assign to one of its top command staff. Whomever is designated to head the logistics sector should be reliable, capable, and knowledgeable.

- Having pre-packaged multiple-casualty incident supplies would eliminate the need to strip ambulances of their medical equipment.

Technical Rescue Considerations

- Rescuers should familiarize themselves with the emergency access techniques outlined in the AMTRAK *Emergency Evacuation Procedures* guide. The MFD quickly discovered that its extrication tools would not work on the train. Several of its power tools were damaged trying to force entry. This is to be expected. In fact, AMTRAK's emergency guide states, "Rail passenger cars and locomotives are constructed to withstand extreme stresses under all conditions. Forced entry is not easily accomplished." Copies of the booklet can be obtained by contacting AMTRAK's Safety Department at (202) 906-4949.

- Highly specialized teams, such as divers, require time and assistance to initiate operations. Involve these teams in pre-incident planning and drills so all department personnel understand how they function, their limitations and capabilities, and what assistance and resources they will require.

Triage and Treatment

- The proper and early alerting of the hospitals proved to be quite helpful. Although area hospitals did not receive the 200+ patients they had prepared to receive, they were ready and had been able to call back additional personnel because they received advanced warning. Be sure to notify the hospitals, in a timely manner, when they are clear to stand down from their disaster plan.

- Distribute patients to avoid overload at any one hospital. Keep the Level 1 trauma facilities available for the most critical patients. The medical control officer ensured that the area's Level 1 trauma center did not receive patients simply because of the nature of the incident. Patients were transported to other area hospitals, leaving trauma beds free for critical trauma patients. This decision, while not easily undertaken or executed, would likely have meant the difference between receiving proper care and not, had critical trauma patients been found at the scene of the derailment.

- After saving the living, it is necessary to have a plan for the removal of the dead. Proper search, documentation, and recovery procedures are crucial for identifying the dead and investigating the cause of the incident. Morgue operations is a continuation of this process. It is advisable to have a mass fatalities procedures annex outlining these procedures in your Emergency Operations Plan. Following a mass fatalities incident, you need to handle the dead properly and address the needs of the family and friends of the victims as well as the needs of the responder. The National Emergency Training Center through the State Emergency Management Training Offices offers the Mass Fatalities

Incident Response Course (G386). Contact your State Training Office for course information.

- If it is a hot day, don't place victims in body bags until they are to be transported. Don't attempt to identify victims on the scene. Morgue personnel should be rotated out as often as possible. Have law enforcement personnel secure the morgue. Keep it sheltered from view and from the weather, if possible.

Handling the Media

- Anticipate media from your area as well as national and even international reporters, if the incident is big enough. Organize and control the media to avoid serious operational problems.

- The Public Information Officer is a critical member of the Command Staff. Without a PIO, the media will seek out anyone willing to talk with them. This could result in them getting incorrect or restricted information. The worst possible circumstance would be that the media hamper the Incident Commander or operational personnel.

- Understand that the media have a constitutional right to report on what is happening. More importantly, they are competitive and value getting a "scoop." If you ignore the media, they will find a way to get the story. It is better to enlist their cooperation by making accommodations for them than to try to keep them in an adversarial position.

- Establish media ground rules early on in the incident. If certain areas are restricted, tell the media where they cannot go and why (i.e., safety considerations, public health, etc.). If news crews violate the rules, have law enforcement escort them off site. If necessary, remind the media that they are being granted special access and that, in return, they need to abide by certain rules.

- A rotation system was used to ensure that all the media got an opportunity to visit the crash site. While this worked well, it may not always be practical. When access to the scene is limited or unsafe, consider the use of a "pool reporter." Under this system, one reporter and one camera operator are selected to go to the site. All the media who are there when the pool is established are then given full rights to any reporting or footage produced by the pool crew.

- Know the facts, and know what information can and cannot be released. Anticipate questions and the flow of events. When the media area is remote, maintain constant contact with the operations site to obtain updated information.

- Consider utilization of a "unified" PIO function (fire, law enforcement, EMS, emergency management, etc.) - it distributes the work and the stress by allowing for rotation of the spokesperson. This will require that PIOs in an area know each other, train together, and use a unified PIO function on normal incidents and in exercises. Have the Incident Commander assign assistants to the PIO.

- When an incident revolves around an organization (such as AMTRAK), have one of its representatives available at the command post to get and give information which only that organization can provide.

VII. Conclusion

The tragic wreck of the *Sunset Limited* presented emergency responders with a complex and unusual set of challenges. It was in many respects, the worst-case scenario come true. As one responder said, "If you had asked me whether I'd ever go to a train derailment in the middle of the night, in the middle of the swamp, I'd have thought you were crazy." It is important to note that trains characteristically travel through remote and inaccessible places. As evidenced by the wreck of the *Sunset Limited* and numerous other train crashes, emergency responders need to anticipate and plan for these types of incidents.

Considering that there was virtually no visibility or land access, that responders had large numbers of patients to triage, treat, and transport, that access to fatalities was technically difficult and dangerous, and that extreme heat made recovery and processing of fatalities an unpleasant and emotionally tortuous task, it is to the credit of all those there that the only responder injury was a cut arm.

By all accounts, training and the use of ICS contributed heavily to the ability of the MFD to handle the obstacles which confronted them in Big Bayou Canot.

VIII. Post Script: Mobile Revisited

Just before 0700 hours on March 20, 1995 (approximately one week after this incident investigation was completed) the MFD responded to a series of motor vehicle collisions on the I-10 Bridge over the Mobile River. Upon arrival units discovered that five separate car crashes involving 193 vehicles had occurred in the fog on the 7-mile-long bridge. Seventy-one people were injured, one fatally. Victims were located on both the east- and west-bound sides of the bridge.

MFD instituted its multiple casualty incident plan. Over 230 emergency personnel from MFD, Baldwin and Mobile Counties, various private ambulance services, and assorted law enforcement agencies responded to the scene. Forty-nine patients with varying degrees of injury were transported by ambulance; transit buses transported the remaining 18 walking wounded patients. Both lanes of the bridge were reopened approximately five and one-half hours after the first crash occurred.

As with the *Sunset Limited* incident, the ICS was used. This time, however, assisting agencies were well-versed in its use. Because of the 800 mega-Hertz radio system, a common radio frequency was available for incident coordination. While two radio channels were used, a third was available had it been necessary to use it.

According to PIO Steve Huffman, "The lessons learned from AMTRAK resulted in us being able to handle this incident without problem." This is perhaps the lasting legacy of the *Sunset Limited.*

* * *

APPENDIX A

APPLICABLE AMTRAK FLOOR PLANS

Class F40PH

A-END 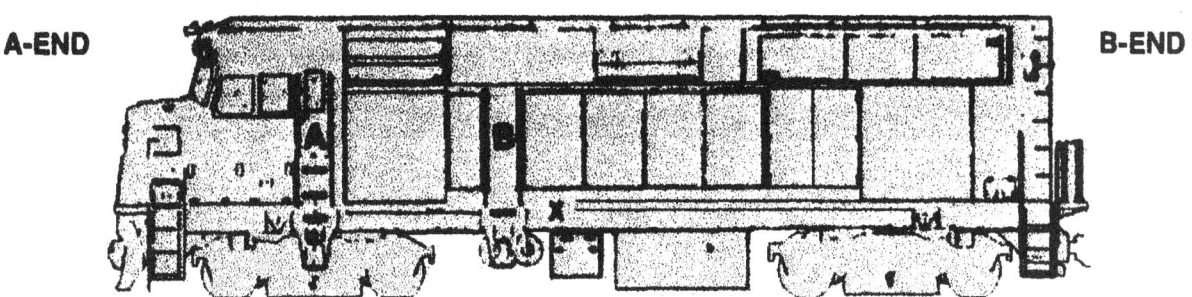 B-END

Model F40PH-3000 H.P. Passenger Locomotive
480 Volt-Head End Power Equipped

A-END 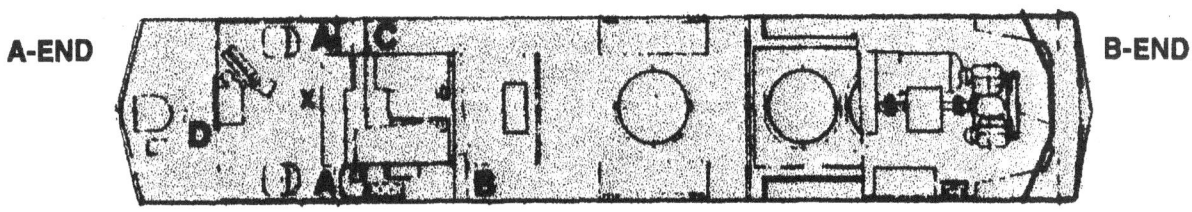 B-END

F40PH–Floor Plan

Baggage Cars

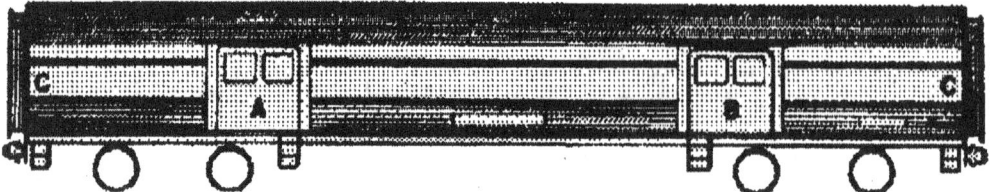

Standard Baggage Car

Floor Plan

Crew Dorm

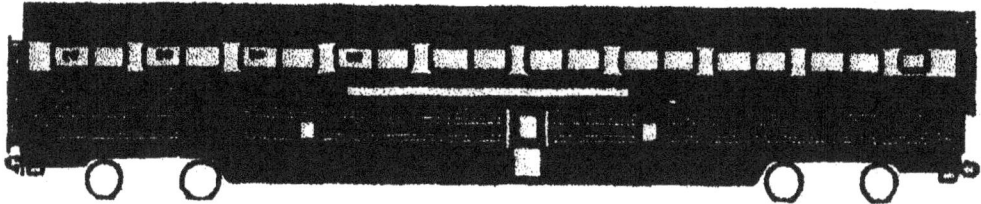

UPPER LEVEL

LOWER LEVEL

Floor Plans

A-END

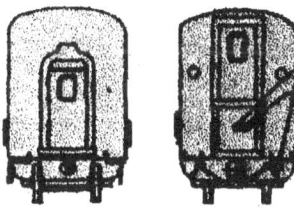

Superliner Lounge

UPPER LEVEL

LOWER LEVEL

Floor Plans

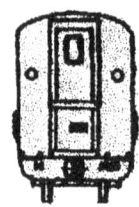

Superliner Coach

LEFT SIDE VIEW

UPPER LEVEL

LOWER LEVEL

Floor Plans

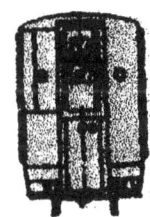

Superliner Diner

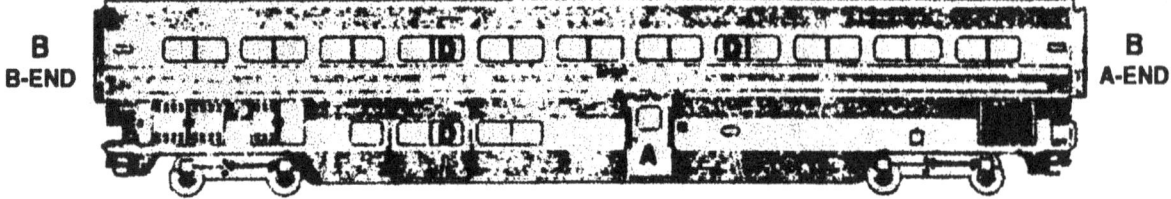

B
B-END

B
A-END

B
B-END

B
A-END

Standard Superliner Diner

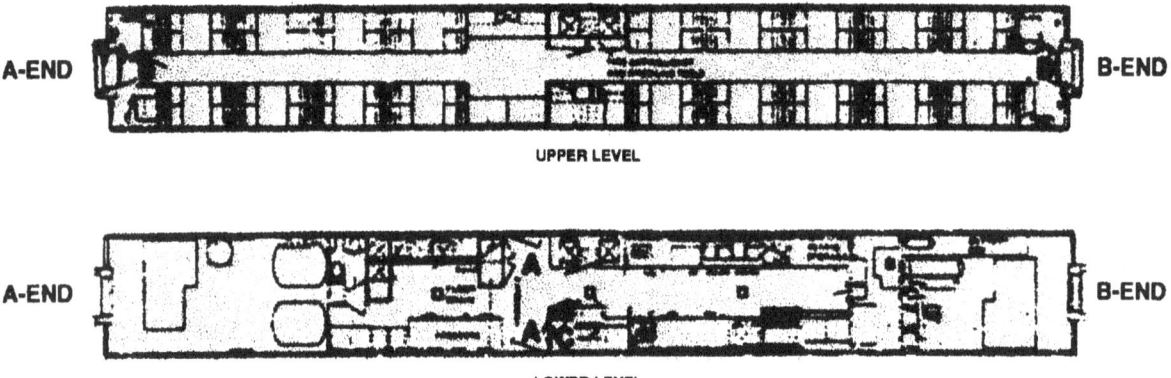

A-END

B-END

UPPER LEVEL

A-END

B-END

LOWER LEVEL

Floor Plans

Superliner Sleeper

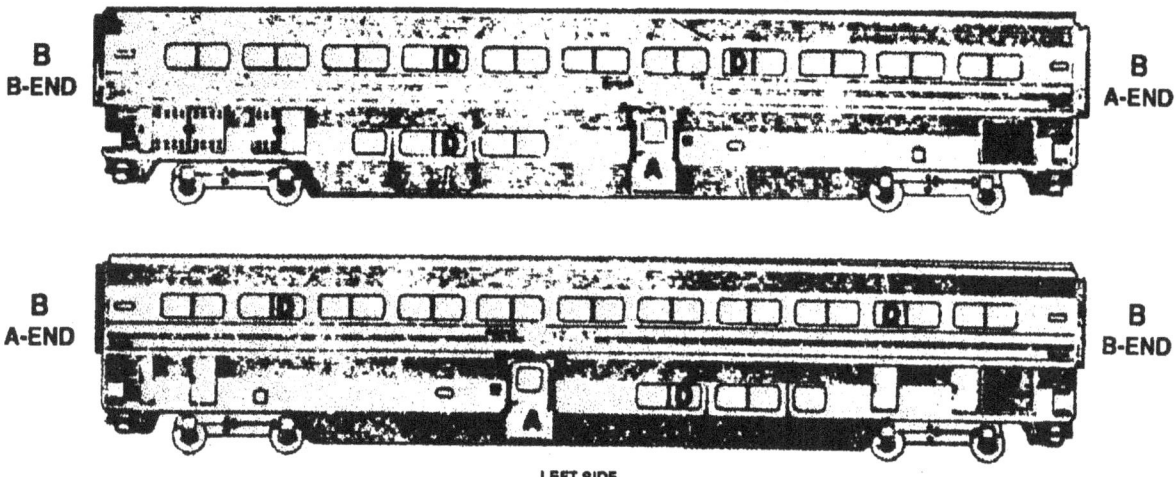

B
B-END

B
A-END

B
A-END

B
B-END

LEFT SIDE

Standard Superliner-Sleeper

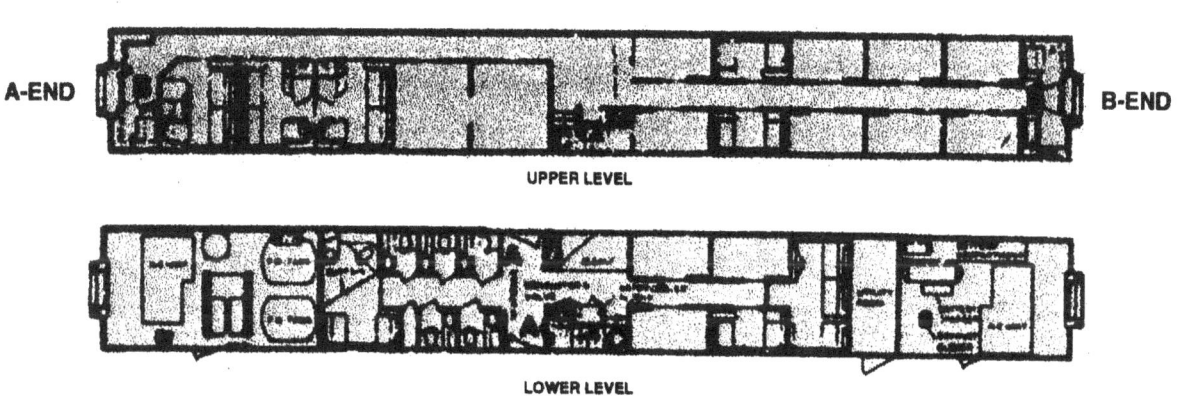

A-END

B-END

UPPER LEVEL

LOWER LEVEL

Floor Plans

◆   ◆   ◆

APPENDIX B

TRANSCRIPT OF 9-1-1 CALLS

9-1-1:   9-1-1

CARR: This is Warren Carr-, the CSX Railroad.

9-1-1:   Yes, sir.

CARR: We got a passenger train at Bayou Sara drawbridge, has derailed. I understand that there's people in the water. The bridge is on fire. I need all.

9-1-1:   Wait a minute 507, where?

CARR: Bayou Sara drawbridge.

9-1-1:   Where is this? Where is that located?

CARR: It's off the Mobile River.

9-1-1:   Um-hm.

CARR: It's north of Chickasabogue draw.

9-1-1:   You can't?

CARR: You can't get there, can't get there by vehicle.

9-1-1:   Okay, is that going to be, is that going to be in Prichard, north of Chickasabogue? Is that going to be it?

CARR: It's north of Chickasabogue.

9-1-1:   Um-hm.

CARR: I'm talking to 9-1-1, Tommy. **(speaking with another CSX employee)**

9-1-1:   Okay.

CARR: It's a passenger train. I got people in the water. I got cars on fire.

9-1-1:	Okay, but . . .
CARR:	It's a derailment.
9-1-1:	Okay, a derailment, but is it in Prichard, there by Chickasabogue Creek?
CARR:	No, it's on the river.
9-1-1:	On the Mobile River?
CARR:	You can't get to it over the road.
9-1-1:	Okay.
CARR:	You're going to have to get some helicopters and boats and Coast Guard and all those people.
9-1-1:	507 drawbridge, on the Mobile River.
CARR:	On CSX Railroad.
9-1-1:	**(unintelligible)**
CARR:	No, it's, it's south of Mobile River, north of Chickasabogue River.
9-1-1:	South of.
CARR:	North of Chickasabogue, at the next, next major creek, north of Chickasabogue.
9-1-1:	And south of, what other river was that?
CARR:	South of Mobile River. It's right along beside Mobile River, where Bayou Sara comes off the Mobile River.
9-1-1:	Bayou Seven?
CARR:	Get me ahold of the Coast Guard.
9-1-1:	Okay, sir. We're going to get someone out there. What's your callback number, sir. You calling from 4-3-4?
CARR:	4-3-4.
9-1-1:	1300?

CARR: No, 1390 or 1375.

9-1-1: What's your name sir?

CARR: Carr, C-A-R-R.

9-1-1: And you're with the C and X railroad?

CARR: That's right. You need all the emergency vehicles that you can get, too.

9-1-1: Okay, you got a derailment with people in the water?

CARR: That's right.

9-1-1: Okay.

CARR: Bayou Sara draw.

9-1-1: All right.

CARR: Bridge, on CSX Railroad.

9-1-1: Okay.

(end of call)

9-1-1: Emergency, 9-1-1.

SEYMOUR: This is Ronnie Seymour, with CSX Transportation.

9-1-1: Uh-huh.

SEYMOUR: AMTRAK has derailed at Bayou Sara drawbridge. I need any available assistance. I need boats in the water. I need them as soon as possible.

9-1-1: Okay, stay on the line just a moment. I've got an operator working on that. Can you hold the line just a moment?

SEYMOUR: Yes.

9-1-1: Just a moment.

9-1-1: Ma'am. **(this is the same 9-1-1 operator from the Carr call)**

UNKNOWN FEMALE: Hold on a minute, please. Yeah.

9-1-1:   Can you tell me what the train?

SEYMOUR: Hello?

9-1-1:   Sir.

SEYMOUR: Yes.

9-1-1:   Can you tell me what the train was carrying?

SEYMOUR: The passengers.

9-1-1:   Passengers? Linda, so it's not a . . .

UNKNOWN FEMALE: It's going to be Saraland, but it's a passenger train that's derailed, **(unintelligible)** crew in water; the train's on fire.

9-1-1:   So it is a passenger train?

SEYMOUR: Yes, ma'am.

9-1-1:   Okay, sir. It's going to be Saraland's, but we have notified the proper authorities, okay?

SEYMOUR: Thank you very much.

9-1-1:   You're welcome, bye-bye.

9-1-1:   Carol, I can't get Marine Police to pick up.

(sound of dial tone)

(end of call)

9-1-1:   9-1-1.

AMTRAK: Yes, listen, this is AMTRAK train, a supervisor. We're on, we're on the Mobile River. The bridge has gone out. We got cars burning, people in the river, can't swim.

9-1-1:	Okay.
AMTRAK:	We need help. Any kind of help you can get down here.
9-1-1:	Okay, sir, just stay on the line with me. Sir?
AMTRAK:	Yes.
9-1-1:	Okay, you're with the AMTRAK train?
AMTRAK:	Yes, ma'am. I'm the supervisor on board. We're on the Mobile River.
9-1-1:	You're on the Mobile River?
AMTRAK:	On the Mobile River. We've got cars burning. They're over the bridge is out. There's people in the water. We're trying to help them, but.
9-1-1:	So the bridge is out?
AMTRAK:	We need all kind of help. Yes, ma'am, we need help, send help, please.
9-1-1:	Okay, just stay on the line, okay? We got someone en route to you. It has been called in before, but I need you to stay on the line, okay? AMTRAK train. Hello, sir?
AMTRAK:	Yes, ma'am?
9-1-1:	Just hold on the line, okay?

(AMTRAK placed on hold, comes back on the line)

AMTRAK:	Location, we can use them.
9-1-1:	Okay, sir?
AMTRAK:	Yes, ma'am.
9-1-1 :	We do have help on the way.
AMTRAK:	Okay.
9-1-1:	Sir, listen to me, what, you, hello, sir?
AMTRAK:	Yes, ma'am.

9-1-1:   This is the bridge that's out?

AMTRAK: Yes, ma'am. The bridge is out. Yes, ma'am.

9-1-1:   Okay, which bridge is this that's, that's out?

AMTRAK: Ma'am. I don't know. We're on the Mobile River.

9-1-1:   You're on the Mobile River?

AMTRAK: That's all I know.

9-1-1:   But you don't know what bridge it is?

AMTRAK: No ma'am. No, ma'am. I haven't been informed by the conductor. John, John?

9-1-1:   Okay, Sir?

AMTRAK: Yes, ma'am.

9-1-1:   Can you give me some information on where this train was coming from?

AMTRAK: Was coming from New Orleans, sir.

9-1-1:   New Orleans,

AMTRAK: Be careful, watch your step. (speaking to passengers) It's left New Orleans. Ma'am, I have to go and assist these folks.

9-1-1:   Okay, sir.

AMTRAK: All right, then.

(end of call)

* * *

APPENDIX C
LESSONS LEARNED FROM THE
AMTRAK/CONRAIL DERAILMENT IN
CHASE, MARYLAND, JANUARY 1987

Reproduced courtesy of the Charles McC. Mathias, Jr.
National Study Center for Trauma and Emergency Medical Systems
© 1989, University of Maryland at Baltimore
All Rights Reserved

Following is a summary of the recommendations made in the post-incident analysis report of the AMTRAK/CONRAIL derailment which occurred January 4, 1987 at Chase, Maryland. Although there are many similarities between this incident and the derailment of the *Sunset Limited,* there are also crucial differences - the two most important being that this incident occurred over land and in cold weather. Sixteen (16) people died; 177 others were injured.

The reader should note that some of the recommendations reproduced below are specific to the Maryland EMS system as it was in 1987, when the derailment occurred. Many of the recommendations have been subsequently implemented. Where appropriate, explanatory notes appear in parentheses.

1. For incidents involving multiple fatalities, the Medical Examiner and his office need to be included in the appropriate notification call list. It was by chance that they were contacted for this incident and were able to respond as they did.

2. Removal of, care for, and discussion pertaining to the deceased or their families should be accomplished with the assistance of a representative of the Medical Examiner's Office, as they are aware of all considerations needed for these operations.

3. Arrangements need to be made for the purchase and storage of disaster pouches and for regional storage locations from where they could be more readily transported to the scene.

4. The purchase of a refrigerated trailer for use as a mobile morgue should be investigated.

5. Accessibility was a severe problem, as there were many checkpoints on access roads and fire and rescue apparatus blocked narrow roads that led to the scene. More easily recognized identification needs to be investigated along with appropriate transportation plans for Medical Examiner's Office to respond to the scene.

6. There should be designated persons from each jurisdiction in the region to respond to the command post to act as a liaison and also control the units from that person's jurisdiction.

7. There needs to be a mechanism to control individuals responding on their own to the scene.

8. Preparation for environmental conditions such as low temperature needs greater attention.

9. Great care needs to be taken in the movement of uninjured passengers. A large group was escorted past the active treatment areas at the scene. Changes in their emotional state could be recognized immediately due to the additional stress of seeing fellow passengers who were injured.

10. "GO teams" (Maryland hospital-based disaster teams) should be used more extensively in triage areas.

11. Staging should be utilized to a greater extent.

12. Hourly updates should be given to hospitals, even if the message is "no new information."

13. EMS Communications System should be improved.

14. Interhospital communications should be established to centralize patient identification/condition list at MIEMSS (statewide communications center).

15. Clearer designation of command personnel is needed.

16. Traffic should be directed well out of the area.

17. Suggestions from civilian experts (i.e., AMTRAK workers) should be considered.

18. There is a need for distinct identification of CIP (Crisis Intervention Preparedness) team members. Clear identification emblems and acronyms on outer garb (vests, jumpsuits, jacket) that specifically identify CIP team members are necessary.

19. Communications on-site between CIP team dyads (pairs working together) and the team leader must be improved. Simple walkie-talkie equipment would solve the problem.

20. The capabilities of organizations like Dogs East (a search dog team) should be considered earlier in future incidents which may involve the possibility of trapped victims.

21. Dogs East normally utilizes a second dog team to confirm the findings of a single team. If the team is to be utilized in the future, arrangements should be made to allow the transport of more than a single team.

* * *

*U.S. GOVERNMENT PRINTING OFFICE: 1996-719495/82752